Bugs

Army Ants

by Trudy Becker

FOCUS READERS®
PIONEER

www.focusreaders.com

Focus Readers is distributed by North Star Editions:
sales@northstareditions.com | 888-417-0195

Produced for Focus Readers by Red Line Editorial.

Photographs ©: iStockphoto, cover, 1, 20; Shutterstock, 4, 6, 8, 10 (top), 10 (bottom), 14, 17; Leonardo Mercon/VW Pics Editorial/AP Images, 12; Francesco Tomasinelli/Science Source, 18

Library of Congress Cataloging-in-Publication Data
Names: Becker, Trudy, author.
Title: Army ants / by Trudy Becker.
Description: Lake Elmo, MN : Focus Readers, [2023] | Series: [Bugs] | Includes index. | Audience: Grades 2-3
Identifiers: LCCN 2022025535 (print) | LCCN 2022025536 (ebook) | ISBN 9781637394472 (hardcover) | ISBN 9781637394847 (paperback) | ISBN 9781637395578 (pdf) | ISBN 9781637395219 (ebook)
Subjects: LCSH: Army ants--Juvenile literature.
Classification: LCC QL568.F7 B37 2023 (print) | LCC QL568.F7 (ebook) | DDC 595.79/6--dc23
LC record available at https://lccn.loc.gov/2022025535
LC ebook record available at https://lccn.loc.gov/2022025536

Printed in the United States of America
Mankato, MN
012023

About the Author

Trudy Becker lives in Minneapolis, Minnesota. She likes exploring new places and loves anything involving books.

Table of Contents

Chapter 1

On the Move

A huge group of army ants is on the move. The ants rush along the forest floor. Other insects run away. But most of them get eaten by the ants.

Army ants live in warm, wet places. But they don't stay put for long. They **raid** new areas. They eat many insects there. Soon, the insects are gone. The army ants move to a new place.

Fun Fact

Some birds follow army ants. They can eat the insects that run away.

Chapter 2

Army Ant Bodies

An army ant has three main body parts. It has a head. It also has a **thorax**. That is the middle part. It has an **abdomen**, too. That is the back part.

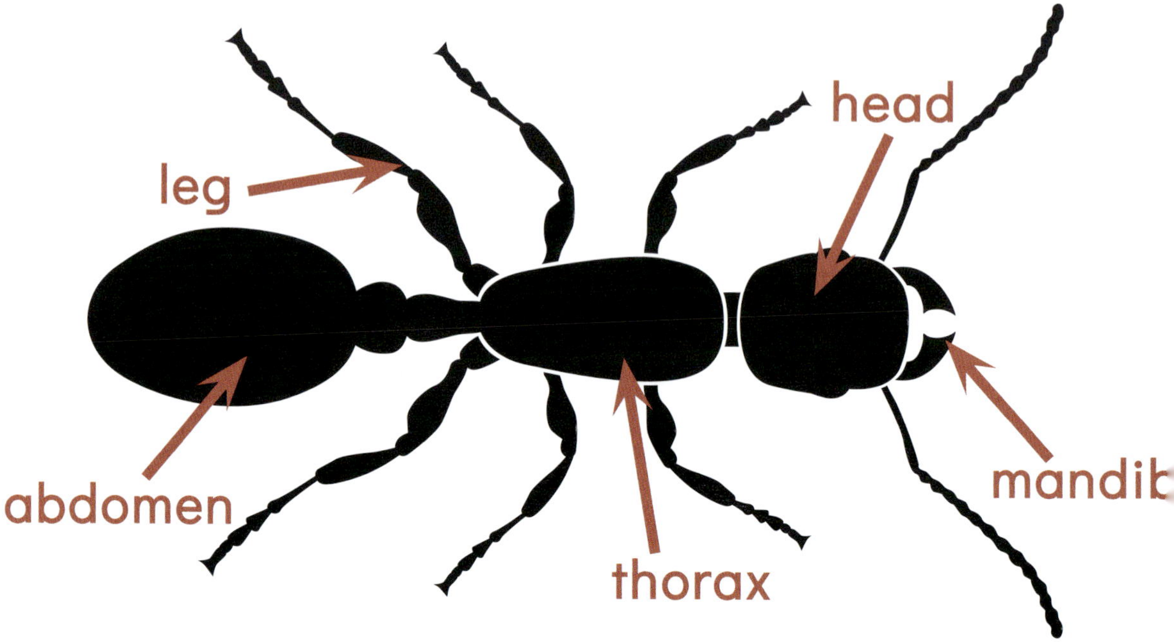

head
leg
abdomen
thorax
mandib

There are different types of army ants. Soldiers have strong **mandibles**. Workers have smaller bodies. Males have wings. And the queen has a big abdomen for laying eggs.

Fun Fact

Soldier army ants use their mandibles to bite enemies.

Chapter 3

Ant Strengths

Army ants use **pheromones**. Pheromones are a type of smell. They send messages. Other army ants sense the pheromones. Then the whole **colony** knows what to do.

Army ants work as a group. They build bridges with their bodies. Many ants link their bodies to cross a gap. Then other ants walk over them.

Fun Fact

Army ant bridges can have up to 50 ants in them.

Making a Home

Some ants live in nests or anthills. But army ants are different. When it is time for the queen to have babies, army ants join together. They use their bodies to make a home. The queen is safe inside.

Chapter 4

Ant Life

Army ants eat insects like wasps and other kinds of ants. But they can eat bigger animals, too. Sometimes spiders or scorpions are their **prey**.

Ants of different sizes have different jobs. Some raid. They bring food to the colony. Many ants take care of the queen. She is busy having babies. When she is ready, the ants move again. It's time to find a new place.

Fun Fact

The queen can lay 300,000 eggs in a few days.

FOCUS ON

Army Ants

Write your answers on a separate piece of paper.

1. Write a sentence that explains the main idea of Chapter 1.
2. What would you do if you saw army ants raid an area?
3. Why does the queen ant have a large abdomen?
   - A. to hold muscles for her strong mandibles
   - B. to store extra food during winter months
   - C. to have a large number of babies
4. Why don't army ants build anthills?
   - A. It's hard to find materials.
   - B. They often move to new areas.
   - C. They are too weak to build.

Answer key on page 24.

Glossary

abdomen
The back part of an insect. It contains the heart and stomach.

colony
A group of animals that live together.

mandibles
Jaws that stick out from an insect's head.

pheromones
Smells that animals make that affect the behavior of others.

prey
Animals that are eaten by other animals.

raid
To attack in a group.

thorax
The middle part of an insect. The legs are attached to this part.

To Learn More

BOOKS

Pallotta, Jerry. *Green Ants vs. Army Ants.* New York: Scholastic, 2019.

Statts, Leo. *Ants.* Minneapolis: Abdo Publishing, 2019.

NOTE TO EDUCATORS

Visit **www.focusreaders.com** to find lesson plans, activities, links, and other resources related to this title.

Index

Answer Key: 1. Answers will vary; **2.** Answers will vary; **3.** C; **4.** B